ESSENTIAL SCIENCE

D0354028

alternative
energy

Marek Walisiewicz

LONDON, NEW YORK, MUNICH,
MELBOURNE, AND DELHI

senior editors Peter Frances and Hazel Richardson
DTP designer Rajen Shah
picture researcher Rose Horridge
illustrator Richard Tibbitts

category publisher Jonathan Metcalf
managing art editor Phil Ormerod

Produced for Dorling Kindersley Limited by
Cobalt id, The Stables, Wood Farm, Deopham Road,
Attleborough, Norfolk NR17 1AJ, UK

First American Edition, 2002
02 03 04 05 10 9 8 7 6 5 4 3 2 1

Published in the United States by
DK Publishing, Inc.
375 Hudson Street
New York, NY 10014

Library of Congress Cataloging-in-Publication Data

Walisiewicz, Marek.
  Alternative energy / Marek Walisiewicz.
    p. cm. -- (Essential science)
  ISBN 0-7894-8919-8 (alk. paper)
  1. Renewable energy sources. I. Title. II. Series.

TJ808 .W35 2002
621.042--dc21

2002071498

Color reproduction by Colourscan, Singapore
Printed and bound by
Graphicom, Italy

See our complete product line at
www.dk.com

contents

energy
addicts

Only 250 years ago, our ancestors relied entirely on natural sources of energy. Animals pulled plows, windmills ground corn, and the principal motive force of society was human muscle. Now, muscle contributes less than one percent to the work done in developed countries, and the vast array of goods and services that we take for granted is supported by our escalating use of fossil fuels – coal, oil, and gas. The prosperity of the industrialized world has been underwritten by fossil fuels, and for decades we have behaved as if these resources would never run out. Today, we are more circumspect. Wars and political crises have shown the fragility of our fuel supplies, and we have all become aware of the environmental impact of our addiction to energy. The age of fossil fuel is coming to an end, and future historians may see it as an anomaly – a time of unsustainable consumption. In the coming decades, we must break our dependence on fossil fuels. The transition may not be easy, and it will have profound consequences for us all.

gas flare
Finding and extracting fossil fuels from the Earth is itself an energy-intensive process. Here, gas is flared off from an oil platform as it is shut down for routine maintenance.

the power balance

resources
This map shows the uneven distribution of known fossil fuel reserves around the world. Mismatches between resources and consumption mean that fossil fuels often have to be transported great distances to their end users.

Humans have an unquenchable thirst for energy. Global demand for power has tripled since 1950, to the point that we now use the energy equivalent of 10,000 million tons of oil every year. According to the World Energy Council, energy consumption is likely to rise by 50 percent by the year 2020. Most of our power comes from fossil fuels – coal, gas, and especially oil, which has become the single most critical resource on the planet.

The deposits of oil on which our economies depend are tens of millions of years old, originating in ancient seas that teemed with microscopic plant and animal life. As these organisms died, their remains settled in seafloor basins, forming a rich organic mud. Over the millennia, the mud was buried and compressed by layers of

oil reserves (50 billion barrels)
gas reserves (200 trillion cubic feet)
coal reserves (40,000 million short tons)
energy consumption (10 quadrillion Btu)

what is energy?

Energy is an elusive concept. It cannot be seen and has no physical substance. We only know it is there because we can see its effects, and we only value it for what it can do for us. Scientists define energy as the capacity to do work – to move something against a resisting force. It is easy to see that a swinging golf club has energy because it has capacity to move a ball down a fairway. This type of energy – the energy of moving things – is known as kinetic energy. Another form of energy is potential energy: the water held back by a dam may not be moving, but it certainly has the potential to do work when it is released. Other types of energy – heat, chemical, electrical, and nuclear – can all be seen as forms of kinetic or potential energy. An object is only hot because its constituent atoms are moving faster than those in a cool object; oil is a store of chemical energy only because its molecular bonds store potential energy; and the electrical energy produced by a generator is a stream of moving electrons within a wire. A basic physical principle called the First Law of Thermodynamics states that energy cannot be created or destroyed. When we "generate" or "use" energy, we are really just converting one form of energy into another. In the process of conversion, some energy is always changed into an unwanted form, so conversion is never 100 percent efficient.

conversion losses
Coal burned in a power station is converted to electricity with an efficiency of about 50 percent; 10 percent more is lost in transmission; and a conventional lightbulb turns electricity into light with an efficiency of just five percent. The overall efficiency using coal to light our homes is therefore a little over two percent.

measuring energy

Scientists measure energy in units called joules (J). One joule is a tiny amount of energy – about one-thousandth of that released by striking a match – so large multiples of millions (MJ), billions (GJ), and trillions (TJ) of Joules express the energy

James Watt
Watts are named after James Watt, inventor of the steam engine.

content of fuels. To confuse matters, different fuels are often measured in different units. One barrel of oil is equivalent to 6.1GJ; one cubic foot of gas is equivalent to 1MJ; one ton of coal is 25GJ; and one British thermal unit (Btu) is 1,055J. Power – the rate at which energy is used or converted – is measured in watts. One watt is equal to one joule per second. Again, large multiples (MW, GW, and TW) are often encountered.

sediment above it, and slowly changed into the complex mix of hydrogen and carbon compounds that we know as oil. Coal and gas have similar prehistoric origins, taking millions of years to form. It is not hard to see that fossil fuels are a finite resource; no matter how efficiently we extract them from the ground, they are sure to run out one day.

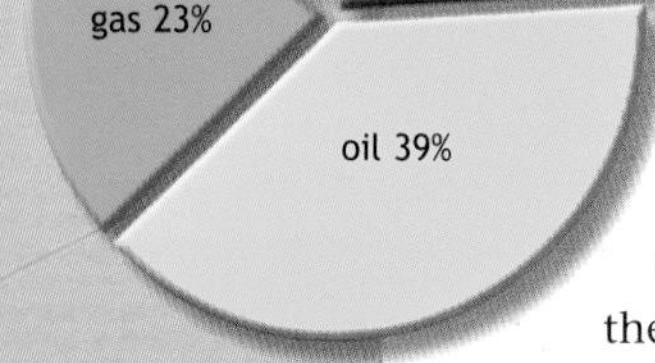

renewables
nuclear   8%
        6%
                    coal 24%
gas 23%
        oil 39%

meeting demand
This chart shows the sources of energy worldwide. The majority comes from fossil fuels.

how much is left?

In 1972, an influential group of planners and scientists published a book, *The Limits to Growth*, that made a set of predictions about the fate of industrialized society. Among other things, the book forecast that world supplies of oil would dry up as early as 1992. Clearly, this has not happened, and this early study highlighted the pitfalls of making such predictions. For a start, it is hard to know exactly how much fossil fuel remains. Whereas the richest coal deposits are typically large seams visible at the surface, oil and gas fields may be hidden several miles below ground. The technology for finding and extracting oil and gas deposits is constantly improving, bringing into production resources that were once considered beyond reach. Furthermore, it is difficult to predict the rate at which known fossil fuel reserves will be consumed in the future. Rates of energy use have a complex relationship with economic conditions and do not always increase predictably over time. In the United States, energy use grew by a staggering 4.5 percent per annum throughout the 1960s, but in the early 1980s it actually fell by 11 percent over a four year period, before rising once again.

Current figures give the "lifetime" of known coal reserves as about 250 years, oil as 40 years, and gas as 70 years, but, in the light of the above cautions, these statistics should be seen as no more than "guesstimates." In fact, most experts agree that outright worldwide shortage of fossil fuels is unlikely to limit economic development over the next few decades.

our energy inheritance

Our current dependence on coal, oil, and gas is an historical legacy. The steam engines that powered the Industrial Revolution in the 18th century needed a

oil saves whales
By the middle of the 19th century, the Right whale was being hunted at a rate of about 15,000 individuals a year. What made it so desirable was the oil in its nose, which made a perfect, clean-burning fuel for lamps and candles. The drilling of the first oil wells, and abundance of mineral oil, probably saved this species from extinction.

strip mining
About 60 percent of the world's coal is produced in open cast mines, where giant coal cutters excavate house-sized bites of coal. This type of mining is likely to become more widespread as oil reserves run low.

concentrated, transportable source of power – coal.
The invention of the internal combustion engine in
the 1870s triggered demand for a liquid fuel with high
energy content – oil. Since then, our whole economic
infrastructure has built up around these cheap, plentiful
fossil fuels, and today's power stations, cars, and even
domestic heating systems are part of this
technological lineage. Many of these technologies
appear antiquated in today's changing energy
climate. Take the car as an example: most of the

power stations

Fossil fuel-based power stations
lie at the heart of industrialized
society. The most efficient designs
burn natural gas, and are capable
of converting 50 percent of its
chemical energy into electricity.
Energy is extracted from the gas in
two stages: first, the gas is mixed
with compressed air and burned
in a combustion chamber. The
hot, expanding exhaust gases rush
through a set of turbine blades,
which spin around, turning a huge
electrical generator. The exhaust
gases, which are still hot, then
enter a boiler where they pass
over pipes containing water,
converting it to steam. The
expanding steam turns another
set of turbines, again linked to
a generator. The electricity
produced is fed into a high-
voltage national grid.

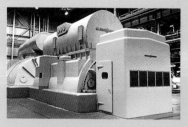

generators
*A power station generator is like a scaled-
up dynamo. Giant magnets in each
generator are spun around at 3,000rpm,
causing an electric current to flow in
surrounding coils of copper wires,
collectively known as the armature.*

cooling towers

chemical energy released when gasoline is burned in its engine is converted not into movement of the pistons, but into heat, which is released as waste. Electric motors provide a far more efficient way of powering cars, yet 99 percent of the world's cars still run on gas. It seems that extravagance is designed into our everyday use of energy, and this is especially so in the world's richest, most developed countries.

Our hunger for energy has led to the centralization of the production and supply of power, putting

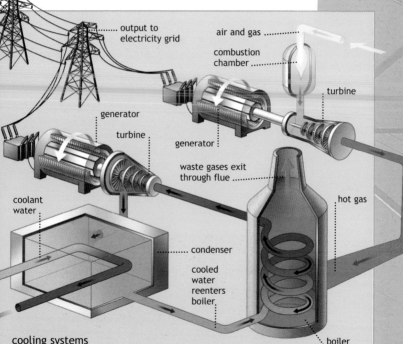

output to electricity grid

air and gas

combustion chamber

turbine

generator

turbine

generator

waste gases exit through flue

coolant water

hot gas

condenser

cooled water reenters boiler

boiler

cooling systems
After the hot steam has passed through the turbines, it is condensed back into water and reenters the boiler to be heated once again. The warmed coolant water is passed into giant steam stacks, where it is sprayed over gravel to be cooled by air entering the towers through their bases.

weighing up
It takes the energy equivalent of 77lb (35kg) of coal to meet the daily needs of every single citizen of the United States. Individual energy consumption in the US is five times the world average.

it into the hands of a few governments and oil companies. Investments of billions of dollars are needed to locate, extract, transport, refine, and distribute fossil fuels, and the reserves themselves are unequally distributed. The Middle East alone contains more than 50 percent of the world's oil in less than 0.5 percent of its land area, and the United States contains over a quarter of the world's coal. Electricity generation is controlled by a relatively small number of power companies, which build and operate thousands of fossil fuel and nuclear power stations worldwide, each producing an average of one billion watts of power.

If the world's reserves of fossil fuels are unlikely to run out in the immediate future, and if it is in the interests of governments and companies to feed our addiction to energy, common sense tells us that there is little impetus for change – that we will continue to burn fossil fuels at ever-increasing rates. But there are new forces at work – economic, environmental, and political – that will force us to change our patterns of energy use in the 21st century. These forces are explored in the next section.

power map
This color-coded image of the Earth at night shows how the use of electricity for lighting is concentrated in industrialized regions (yellow). The red parts of the image correspond to oil flares.

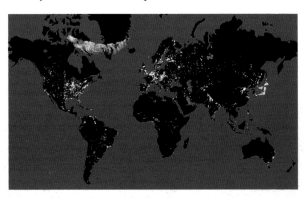

limits to growth

Until the 1950s, the United States was entirely self-sufficient in oil, pumping enough "black gold" from the giant oilfields of Texas and Louisiana to meet domestic demand. Today, the country relies on imports from the Middle East, Central America, and Nigeria for about half of its oil supply. These imports make the US highly vulnerable to external political events: for example, in 1973, the Arab members of OPEC (a cartel of the Oil Producing and Exporting Countries) imposed an oil embargo against selected Western countries. Fuel prices tripled overnight, sending shock waves through the entire US economy; and similar economic turmoil was witnessed during the Iranian revolution, and again in the Gulf War. These fuel crises prompted a scramble of investment in energy conservation and renewable power, but their effects were short-lived. Once the imminent dangers had passed, politicians soon bowed once again to the pressures

"On our crowded planet there are no longer any internal affairs."

Aleksandr Solzhenitsyn, Nobel laureate

crowded world
Increases in the world's population over the coming decades will transform the politics and economics of energy supply and production.

exploration and extraction

detonator

detectors

sound source
(explosive charge)

sound waves reflected
from rock boundaries

Over millions of years, high temperatures and pressures below ground transform the remains of once-living organisms into coal, oil, and gas. Oil and gas are fluid, so they can move through porous rocks away from their original source. They often become trapped beneath impermeable caps of mudstone or gypsum. The largest of these underground reservoirs can hold more than 500 million barrels of oil. Likely sites of these deposits are located by a range of geological techniques

seismic survey
The sound of a detonation at the surface is reflected back up by the boundaries of different rock layers. Detecting these reflections lets geologists draw up maps of buried rock formations to find oil and coal.

before test drilling begins. Once a viable well has been sunk, the sides of the borehole are lined with a protective coating and valves are installed at the top to control the flow of oil and gas.

trapped riches
Oil and gas are typically trapped in a cup-shaped fold of impermeable rock called an anticline. If they do not rise up the well under their own pressure, water or steam may be injected to force them to the surface.

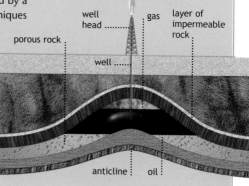

well head

gas

layer of impermeable rock

porous rock

well

anticline

oil

of economic growth fueled by coal, oil, and gas. Political instability in the world's oil- and gas-rich regions returned to haunt the West in the 21st century and many people now question the costs of controlling our energy supply through diplomacy and military intervention.

escalating costs

Back in the 1860s, oil was a plentiful and unexploited resource. Early prospectors only needed to sink shallow wells in order to hit large reservoirs of oil that gushed to the surface under their own pressure. The days of gushers are over. The most easily accessible reserves of oil and gas have been exhausted, and today's oil companies need to work hard to find new underground deposits. The average oil well is now more than 2 miles (3km) deep, and only about one-third of new wells actually hit oil. Finding oil has become a costly activity that takes today's prospectors into the most extreme environments – remote deserts, the frozen Arctic, and, most notably, underwater. The seas off the Arabian Gulf, North Sea, and Gulf of Mexico now provide about one-third of the world's oil. Bringing these far-flung resources into production is even more expensive: the Statfjörd B oil platform, which towers 890ft (270m) over the North Sea, is one of the largest and most costly structures on the face of the planet. Transporting oil over thousands of miles from well to refinery adds still more to the bill.

> **"No matter how advanced our economy might be, no matter how sophisticated our equipment becomes, for the forseeable future we will still depend on fossil fuels."**
>
> George W. Bush, 2000

oil fire
Maintaining fossil fuel supplies exacts a high financial and environmental toll from the developed world. Destruction of oil wells during the 1991 Gulf War released an estimated six million barrels of oil into the environment of Kuwait every day.

The economics of oil, and, to a lesser extent, gas production are becoming marginal, and there will come a point when the energy obtained from a well will be matched by the energy put in. Long before that happens, other energy technologies – wind, water, solar, biomass, and geothermal – will become economically attractive, and are likely to be adopted on a large scale in the industrialized world.

environmental limits

Political and economic forces will rein in our use of fossil fuels in the coming decades, but the biggest shift in our patterns of energy use will be prompted by a more urgent concern – damage to the environment.

Whenever fossil fuel is burned in cars and power stations, carbon dioxide (CO_2) gas is released as a waste product. This gas occurs naturally in our atmosphere, where it acts like a blanket, trapping heat from the Sun and warming the Earth's surface. Indeed, without this "greenhouse effect," our planet would freeze over. But many scientists now believe that human activity is releasing

winds of change
Climate change, caused in part by our use of fossil fuels, has brought about shifts in global weather patterns. In the coming years, extra heat energy trapped in the atmosphere is likely to make tropical cyclones ever more violent and increase the risk of flooding.

so much CO_2 into the atmosphere that the global greenhouse is getting hotter: the UN's authoritative Intergovernmental Panel on Climate Change (IPCC) forecasts that surface temperatures will rise by 2.7–6.3°F (1.5–3.5°C) by 2100. This may not seem like a lot, but it may wreak havoc in many parts of the world. As the planet heats up, weather systems will change dramatically: some countries will become wetter, others drier; some will be subject to more violent storms and floods, while others may actually become cooler. There is a theory, for example, that global warming could switch off or deflect the course of the Gulf Stream – the current that brings warm water across the Atlantic – turning Western Europe into a freezing, dry desert. If the more pessimistic models are correct, global warming could bring about sea level rises of an estimated 3ft (1m) within 50 years. Countries such as the Maldives would simply disappear under the ocean, and millions of people would be displaced from low-lying river deltas, including the Ganges and Mekong.

Global warming is only part of the story: burning coal in power stations produces waste gases that combine with water in the atmosphere to form clouds of sulfuric and nitric acids. The acidic clouds can be carried far away from the source of pollution before dumping their load as rain.

it's a gas
Experts estimate that in the US alone, the environmental costs of gasoline use run to between $200 billion and $900 billion per year.

pipe dreams
Transporting fossil fuels is a major undertaking. The largest oil field in the US, on the coast of Alaska 250 miles (400km) north of the Arctic Circle, is linked to the port of Valdez by a 625 mile- (1,000km-) long pipeline that crosses three mountain ranges and three major earthquake zones.

This "acid rain" causes direct damage to plant leaves and kills fish in freshwater lakes. Despite attempts to regulate and clean up the output of power stations, acid rain is responsible for massive damage to ecosystems in the northeastern US, Canada, and Scandinavia.

Transporting fossil fuels around the world also exacts a high environmental toll. Today's largest oil tankers have capacities in excess of 400,000 tons, and the world's fleet has a very good safety record. But on the rare occasion when something goes wrong, it can cause catastrophic damage. When the tanker *Exxon Valdez* hit a reef on March 24, 1989, 11 million gallons (42 million liters) of crude oil poured into the wilderness area of Prince William Sound, Alaska. About 1,250 miles (2,000 km) of shoreline was coated in a thick sludge, local wildlife was decimated, and the cleanup operation involved an army of 11,000 volunteers. There have been more than 50 oil spills of a similar scale since 1970, and smaller quantities of oil are also discharged routinely from ships and leaking pipelines, with the result that today there are virtually no coastal areas left untainted by fossil fuel pollution.

Over the last two decades, the environmental movement has brought the hidden costs of fossil fuel use to the world's attention. Many governments, however, are still not really paying attention. At the 1992 Earth Summit in Rio de Janeiro, the developed nations pledged to keep levels of greenhouse gas emissions in the year 2000 to the same levels they had been in 1990. They failed. Economic concerns won the day, and emissions in 2001 were about 14 percent higher than the Rio target.

acid rain
Forests are damaged by acid rain, which causes lesions on green leaves and leaches vital nutrients, such as calcium and magnesium, from the soil.

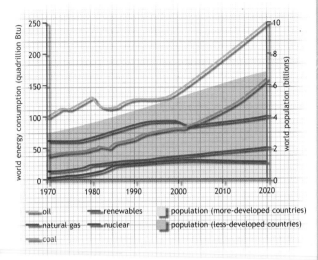

legend:
- oil
- natural gas
- coal
- renewables
- nuclear
- population (more-developed countries)
- population (less-developed countries)

future prospects
Past and projected patterns of energy consumption are combined with estimates of population growth in this graph. These figures from the US Energy Information Administration suggest that our use of oil and coal will rise steeply in the next decades.

In just one generation, the human population of our planet has increased by 33 percent, and is now more than six billion. The bulk of this increase has taken place in the developing world, where the average annual consumption of energy is 0.8 tons of oil per person, compared with 4.8 in industrialized countries. But the less-developed countries are fast catching up with the West; in China and India, electricity consumption is rising by seven percent a year, and best estimates suggest that within 20 years, developing countries will contribute 44 percent of the world's total CO_2 emissions – a significant increase on today's figure of 28 percent.

We are sitting on a global environmental crisis in the making – a crisis that demands a response from government and industry. In the short term, technical fixes may buy us some time. Judicious use of nuclear power, energy conservation, and strict controls on emissions will extend the shelf life of our existing fossil fuel reserves; but in the longer term, renewable energy is sure to play an increasing role in meeting our needs.

a technical fix?

I t is tempting to trust in technology to solve our looming energy crisis. Surely cutting-edge science will – one day – allow us to generate plentiful, cheap, and clean power, freeing us from our addiction to fossil fuels? But the prospects for such a technical fix are not good. When nuclear fission was first discovered, and the first commercial civil reactor came onstream in 1957 in Shippingport, Pennsylvania, many people believed that the Holy Grail of energy had been found. Yet fission energy has fallen short of the world's expectations. Nuclear fusion, too, has been touted as our ultimate energy source, but today's scientists are not sure that the technology can ever be made to work economically, and research in the US has recently been scaled back by 40 percent. In the meantime, less ambitious technologies are making a real difference to the energy equation; simple, often unglamourous, devices and building techniques are helping us to use power with ever-increasing efficiency, making best use of our existing resources.

atomic energy
Rods containing pellets of nuclear fuel are lowered into the reaction chamber of a power station. Nuclear energy remains important in meeting global demand for power, although it faces an uncertain future.

nuclear power, a failed promise?

first split
The New Zealand physicist Ernest Rutherford is the founder of modern atomic theory. His exhaustive studies on the nature of radioactivity in the early 20th century enabled scientists to split the atom.

On September 16, 1954, Lewis Strauss, the Chairman of the US Atomic Energy Commission, stood before an audience of science writers in New York and confidently declared that their children would enjoy electricity "too cheap to meter." The promise of unlimited nuclear power – a peacetime dividend of atom bomb research – was music to the ears of a world emerging from the austerity of war. Yet in half a century, the promise has not been fulfilled, and nuclear energy has fallen from grace following concerns over its cost, safety, and the intractable problems posed by the disposal of its waste products. In the 1970s, the high cost of electricity from nuclear power stations prompted US utilities to cancel 121 planned reactors. Then, in 1979, a narrowly averted disaster at the Three Mile Island reactor in Pennsylvania

power plants
The dome of this power station houses the radioactive core of a nuclear reactor. Two pounds (1kg) of the nuclear fuel uranium can yield as much energy as burning 12,000 barrels of oil.

tainted public perceptions of nuclear energy for a decade, followed by the accident at Chernobyl, which violently renewed the world's concerns.

In economic terms, too, nuclear energy has been at best a disappointment and at worst – in the words of *Forbes* magazine – "a disaster of monumental scale." The last commercial reactor to be completed in the US took almost 23 years to build at an overall cost of more than $7 billion. Hardly too cheap to meter.

Not surprisingly, many countries in Western Europe have frozen new construction of nuclear power stations, with Sweden and Germany actually planning to phase out exising plants. In the US, it is likely that half of today's nuclear capacity will vanish by 2020 as existing power stations reach the end of their 40-year lifespans. It is too early, however, to write the industry's obituary. There are still 440 reactors in operation around the world, and countries such as France and South Korea depend on "nukes" for over half of their electricity supply. New nuclear power stations are the cornerstone of Japan's ambitions for energy independence, and China plans to quadruple its nuclear capacity in the next two decades.

Even in the West, there are many voices calling for a greater reliance on nuclear energy. They have a case: nukes do not produce the greenhouse gas carbon dioxide, so contribute less to global warming. Huge improvements in reactor safety and efficiency have been achieved over the last few years and many governments are now

> "Since I do not foresee that atomic energy is to be a great boon for a long time, I have to say that for the present it is a menace."
>
> Albert Einstein, 1945

the china syndrome
The ultimate nuclear nightmare is meltdown – when the reaction in the core runs out of control. In the early days of nuclear energy, scaremongers in the US came up with the implausible idea of a core melting its way right through the Earth to emerge in China.

funding new technologies that they believe will rehabilitate nuclear power in the 21st century.

Today's nuclear power stations operate in much the same way as fossil-fuel plants (see p.10). Heat energy is used to pressurize a gas, which in turn drives turbines linked to electrical generators. The difference with nuclear power is that the heat comes from the fission (splitting) of unstable nuclei of naturally radioactive elements, rather than from burning gas or coal. The fuel most commonly used is an isotope (form) of uranium, called uranium–235

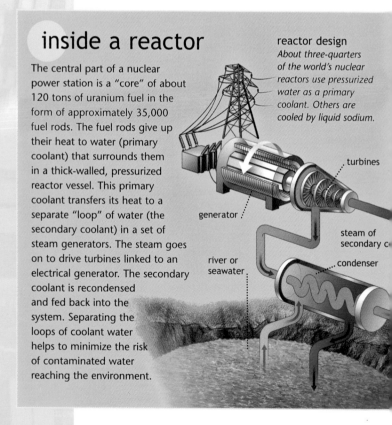

inside a reactor

The central part of a nuclear power station is a "core" of about 120 tons of uranium fuel in the form of approximately 35,000 fuel rods. The fuel rods give up their heat to water (primary coolant) that surrounds them in a thick-walled, pressurized reactor vessel. This primary coolant transfers its heat to a separate "loop" of water (the secondary coolant) in a set of steam generators. The steam goes on to drive turbines linked to an electrical generator. The secondary coolant is recondensed and fed back into the system. Separating the loops of coolant water helps to minimize the risk of contaminated water reaching the environment.

reactor design
About three-quarters of the world's nuclear reactors use pressurized water as a primary coolant. Others are cooled by liquid sodium.

turbines

generator

river or seawater

steam of secondary c

condenser

(U-235), which is mined worldwide as the ore pitchblende. It is purified, concentrated, and compacted into pellets before being sealed into long rods, which form the fuel that is introduced into a nuclear reactor.

U-235 is radioactive. Its nuclei split apart spontaneously into two smaller atoms, releasing heat as well as two or three fast neutrons. If these neutrons then collide with adjacent U-235 nuclei, they cause them to split apart too, releasing more heat and yet more neutrons. If there is enough U-235 present (about 9lb, or 4kg, will do), a

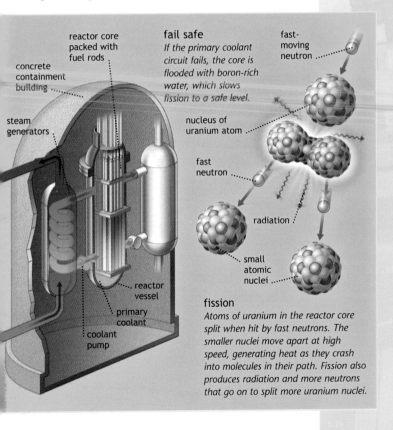

reactor core packed with fuel rods

concrete containment building

steam generators

reactor vessel

primary coolant

coolant pump

fail safe
If the primary coolant circuit fails, the core is flooded with boron-rich water, which slows fission to a safe level.

fast-moving neutron

nucleus of uranium atom

fast neutron

radiation

small atomic nuclei

fission
Atoms of uranium in the reactor core split when hit by fast neutrons. The smaller nuclei move apart at high speed, generating heat as they crash into molecules in their path. Fission also produces radiation and more neutrons that go on to split more uranium nuclei.

chain reaction starts that releases enormous amounts of energy: this is what gives a nuclear weapon its destructive power. In a reactor, the rate of fission is closely controlled by using graphite rods to absorb the excess fast neutrons, thus producing a steady source of heat energy, which is transferred to a coolant – usually water pressurized to about 150 atmospheres.

Today's reactors are behemoths, each producing about 1,000MW of power. Their core of fuel rods is about 13ft (4m) in diameter and is contained in a steel vessel with walls 8in (20cm) thick to withstand the huge pressure of the coolant: about 200 complex subsystems are needed to keep each station running smoothly. But designs in the future will be very different. A coalition of industrialized countries is already developing the next generation of reactors, which will be smaller (about 100MW), simpler (composed of only 25 subsystems), cheaper, and inherently much safer than today's monsters. A prototype of this new design is likely to be onstream by 2006, and may indeed mark the renaissance of the nuclear industry.

domino effect
Nuclear fission and fusion both proceed as chain reactions. They need to be closely moderated in order to release heat energy in a controlled manner.

fusing nuclei
The raw materials of nuclear fusion are two types, or isotopes, of hydrogen – tritium and deuterium. Extreme temperature and pressure induces the nuclei to fuse, producing helium, a free neutron, heat, and radiation.

radiation

neutron

deuterium nucleus

heat energy

tritium nucleus

helium nucleus

radiation

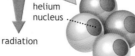

- ○ proton
- ○ neutron
- ○ tritium
- ● deuterium
- ● helium

No matter how efficient it is, fission will always have one big drawback: waste. A reactor produces 20 tons of spent fuel every year, which remains dangerously radioactive for more than 10,000

years. At present, 50 years' worth of this high-level waste is stored "temporarily" in water-filled cooling ponds, but finding it a permanent home is fraught with problems. One possible burial site at Yucca Mountain, near Las Vegas, has been under investigation for 20 years at a cost of $7 billion; a decision on its suitability has not been made.

The only long-term hope for waste-free nuclear power is fusion – the energy source that powers the Sun and stars. In fusion, the nuclei of hydrogen atoms are forced together to produce helium and vast amounts of energy in a self-sustaining reaction. In theory, just one ounce (25g) of raw material can provide a lifetime's energy for an individual in an industrialized country. But there is a hitch. Forcing the nuclei together requires them to be

heated to more than 90 million °F (50 million °C) to overcome the electrical forces that normally keep them apart – a little like pushing together the "wrong" ends of two giant magnets. Despite the investment of billions of dollars in fusion research since the 1950s, scientists have managed to sustain a fusion reaction for only a matter of seconds.

cold fusion
In 1989, Stanley Pons and Martin Fleischmann, two scientists at the University of Utah, made headlines around the world when they claimed to have achieved cold fusion in a simple tabletop apparatus. However, other experimenters failed to replicate their work, and most of the scientific community no longer considers cold fusion a real phenomenon.

fusion reactor
This prototype fusion reactor takes the form of a giant doughnut that encloses the hot hydrogen raw materials. These are in the form of plasma – a mixture of positively charged nuclei and negatively charged electrons, which is controlled by strong electromagnets.

leaner and cleaner

Spending billions of dollars on research into novel energy technologies is not the only way to satisfy future demand. By using power thoughtfully, and reevaluating the whole chain of energy supply – from oil well and coal face to the electrical sockets in our homes and the fuel tanks in our cars – it is possible to make savings of up to 90 percent in the amount of energy we use. Conservation offers a tangible and immediate means of lessening the effects of fossil fuel use, but it requires that we all adapt our lifestyles to some extent.

Increasing energy efficiency does not necessarily mean putting the brakes on economic growth. In the US, for example, individual wealth has increased by 40 percent since 1980, while per capita energy consumption has

save it
Energy conservation may be short on glamour, but gradually changing the way we use energy may pay more dividends than succumbing to the quick fix of new technology.

Trombe wall
Simple changes in building design can help cut the amount of energy used for domestic heating. The Trombe wall – a thin, glazed air space in front of a heat-storing wall – makes elegant use of free solar energy.

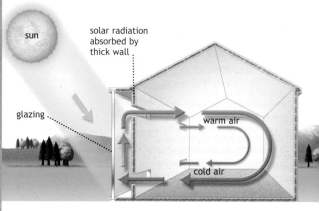

sun

solar radiation absorbed by thick wall

glazing

warm air

cold air

green geometry

Carefully considering a building's shape can make significant savings in the amount of energy needed for heating. Older office buildings and factories were constructed to maximize the penetration of natural daylight. With a large surface area for their volume, they lost heat quickly. Today, architects favor "deep plan" designs that retain heat. The cost of the additional lighting required is small compared to that saved on space heating.

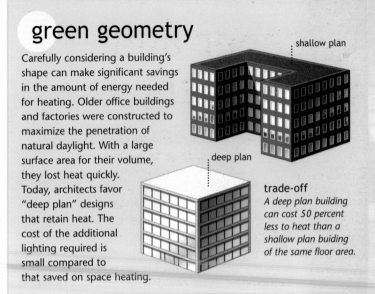

shallow plan

deep plan

trade-off
A deep plan building can cost 50 percent less to heat than a shallow plan buiding of the same floor area.

actually fallen by three percent in the same period. All around the world, new innovations in architecture and transport, as well as in the design of power stations and electrical appliances, are helping to save energy and offset increasing global demand for fossil fuels.

on the home front

About 32 percent of the energy output of the developed world is used in transportation; 25 percent powers industry; and more than 40 percent is supplied to our homes and offices. Most domestic energy is needed for space and water heating, and it is not surprising that some of the biggest energy conservation gains have been made in this area. Research in Sweden and Canada has given us "superinsulated" houses, which employ ingenious construction methods to minimize energy loss. The walls sandwich a 8in (20cm) layer of insulating mineral wool between two layers of lightweight concrete; the windows

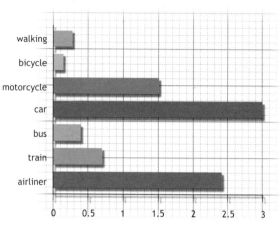

megajoules (MJ) per passenger kilometer (0.6 miles)

are triple glazed; and a controlled ventilation system uses outgoing air to preheat fresh air intake, thus retaining 70 percent of the heat from the "stale" air. Some of the superinsulated buildings have such low heating requirements that the warmth given off by lighting, cooking, and the bodies of the occupants is enough to keep them at a comfortable temperature all year round.

Other buildings – known as passive solar designs – are specially constructed to trap and retain sunlight energy to cut down on heating bills. Typically, they have large glazed areas that face toward the Sun, coupled with thick walls and floors that store up heat during the day and release it during the night. They may be partially buried in the ground: just 10ft (3m) below the surface, seasonal temperature variation is a few degrees, whereas outside it may range beween 86°F (30°C) and 5°F (–15°C). Using the Earth itself as a heat buffer can make a great contribution to energy conservation, and tens of thousands of "earth sheltered" buildings have been built in the US since the 1970s.

In an industrialized nation like the UK, private cars account for a staggering 80 percent of all the energy used for transportation, and about 25 percent of the nation's output of carbon dioxide. Plenty of research has gone into improving vehicle efficiency, with features such as lower drag coefficients, lightweight materials, and computer-controlled engines all contributing toward improved fuel economy in today's mass-produced cars. The development of "zero-emission" electric automobiles is still hampered by the problem of power storage in electric batteries. Conventional lead–acid batteries weigh more than 500lb (230kg), take about eight hours to charge, and give a range of only 200 miles (320km); while more advanced lithium or sodium–sulfur batteries are very costly and have short lifespans. However, cars that run on a combination of gasoline and batteries have a more promising future. In these "hybrids," a lightweight gas engine is supplemented by an electric motor powered by a small battery pack. Whenever the car brakes, the motor acts in reverse – as a generator – refilling the battery, which never needs external charging. This ingenious piece of engineering has already been used in family cars, allowing them to achieve fuel consumption in excess of 70mpg. The reductions in both energy use and atmospheric pollution are substantial.

passive benefits
This building teams energy efficiency with striking architecture. A huge glazed surface maximizes the heating effect of sunlight.

driving force
A typical household in the US spends nearly 20 percent of its income in driving costs – more than it spends on food. Best estimates suggest that nine billion gallons of fuel are wasted in traffic congestion each year.

renewable
resources

Energy is all around us. The interior of the Earth is a cauldron of molten rock, and our planet's atmosphere is bathed in vast amounts of solar energy – enough to satisfy current world demand 15,000 times over. Despite this huge reservoir of natural, renewable energy, we continue to rely on fossil fuels and nuclear power for reasons of convenience, cost, and political expediency. A major problem faced by advocates of renewable energy is the difficulty of harnessing such diffuse resources on a large enough scale. For example, sunlight reaching the Earth's surface is spread over a vast area, making it difficult and expensive to collect. But in recent decades – spurred on by environmental concerns and regular fuel crises – engineers have devised technologies that make renewable energy competitive. The fact that many of these engineers are employed by companies that built their fortunes on oil and gas seems a very reliable indicator that renewable energy will become an important part of all our lives in the 21st century.

sun trackers
*Rows of reflectors
at a solar power
station in
Albuquerque, New
Mexico, concentrate
the light of the Sun
onto a giant boiler.*

watts from water

Water power is at once the oldest and the most highly developed of all renewable energy technologies. Three thousand years ago, water wheels provided people with their first real alternative to muscle power, driving mills to grind corn and moving water in irrigation systems. These creaking wooden devices, which were developed independently in different parts of the Near and Far East, were the forebears of today's giant, advanced hydroelectric installations, which generate 19 percent of the world's electricity at operating efficiencies of up to 90 percent.

Hydroelectric power stations work on simple principles. Turbines extract energy from moving water as it flows down a river and use this energy to turn electrical generators. Damming the river gives far more control over the amount of water flowing through the turbines, so the output of the power station can be matched to demand. The engineering challenge of hydroelectric power is one

hoover dam
At 725ft (221m), the Hoover dam on the Colorado River in Nevada is the highest in the United States. Generators within the dam convert the potential energy of the stored water into 1,800MW of electrical power.

go with the flow

Hydroelectric stations can produce anywhere from a few hundred watts to more than 10,000MW of power. The output of any one installation depends on both the volume of water flowing through the turbine and on the "head" – the vertical distance from the turbine to the surface of the water of the lake or reservoir above. In a high head plant, water may fall perhaps 660ft (200m) from the reservoir through a channel, or penstock, reaching a pressure equivalent to 20 atmospheres by the time it reaches the turbine. A relatively small flow of this fast-moving water is enough to produce significant power output. In a low head plant, the water may fall through just a few yards; a large volume of water, and correspondingly larger turbines, are needed to extract useful amounts of power.

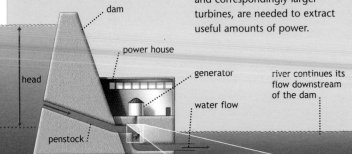

dam

power house

generator

river continues its flow downstream of the dam

head

water flow

penstock

medium head plant
A small-scale installation with a head of about 100ft (30m), such as that shown above, may produce about 500kW from a water flow of 100ft³ (3m³) per second.

water turbine
The turbines used in a medium-head hydroelectric plant resemble ships' propellers and are about 3ft (1m) across. Fixed vanes above the propeller smooth out the flow of water.

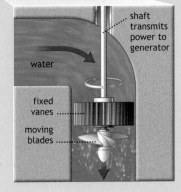

shaft transmits power to generator

water

fixed vanes

moving blades

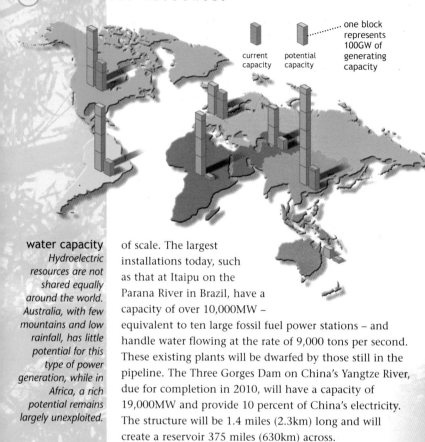

one block represents 100GW of generating capacity

current capacity   potential capacity

water capacity
Hydroelectric resources are not shared equally around the world. Australia, with few mountains and low rainfall, has little potential for this type of power generation, while in Africa, a rich potential remains largely unexploited.

of scale. The largest installations today, such as that at Itaipu on the Parana River in Brazil, have a capacity of over 10,000MW – equivalent to ten large fossil fuel power stations – and handle water flowing at the rate of 9,000 tons per second. These existing plants will be dwarfed by those still in the pipeline. The Three Gorges Dam on China's Yangtze River, due for completion in 2010, will have a capacity of 19,000MW and provide 10 percent of China's electricity. The structure will be 1.4 miles (2.3km) long and will create a reservoir 375 miles (630km) across.

hydro futures

Although hydroelectric power is clean, producing no emissions during operation, the vast development needed has clear drawbacks. The giant reservoir created by the Yangtze scheme will displace more than one million people, submerge 100 towns, and destroy many valuable habitats. Dams can concentrate pollutants that enter the water from large cities upstream, and disrupt important activities downstream; construction of Egypt's Aswan dam in 1964, for example, seriously disrupted fish stocks and the fishing industry in the Eastern Mediterranean.

Hydroelectric generation has massive undeveloped potential: the current world capacity of about 700GW is a small fraction of the estimated 3TW that could be generated if all accessible resources were used. At present, the use of hydroelectric generation is increasing worldwide, but only at a relatively modest rate of 1.5 percent a year. Its adoption is being slowed by concerns about the economic viability and environmental effects of building yet more massive dams and reservoirs.

The growing exploitation of hydroelectric power on a smaller scale (so called micro-hydro) may yet breathe new life into this old technology. Micro-hydro plants have a capacity of less than 5MW – ideal for supplying local villages and industries without the costs of long-distance power transmission. They have minimal impact on the landscape and can sometimes even be "bolted on" to existing water works and sluices, thus minimizing construction work. Micro-hydro is burgeoning in countries like China, where more than 100,000 units have been installed, and the technology is rapidly extending its reach across the world.

wave power

Hydroelectric generation is really just a means of tapping into second-hand solar energy. It is heat from the Sun that drives the Earth's weather systems, vaporizing water, which then falls as rain to feed streams and rivers. The same goes for another renewable technology based on moving water – wave power. Waves are born when wind stirs up the surface of open ocean and creates local storm waves. Their turbulent energy dissipates and is carried

surf's up
The raw power of the oceans is well known to surfers, but wave energy may one day help to meet demand for clean electricity, especially in the wave-battered countries of the North Atlantic.

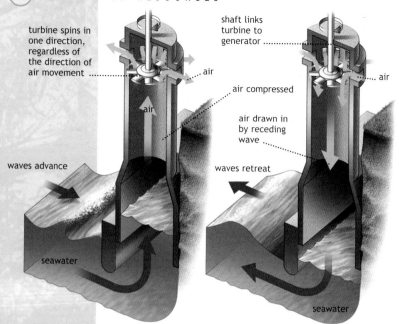

turbine spins in one direction, regardless of the direction of air movement

shaft links turbine to generator

air

air

air compressed

air

air drawn in by receding wave

waves advance

waves retreat

seawater

seawater

shore thing

In this wave energy device, an air chamber rises above the level of the water. Approaching wave crests compress the trapped air, forcing it through a turbine coupled to a generator. As the wave recedes, air drawn in spins the turbine again.

away – over distances of up to thousands of miles – in the form of smoother swell waves. As the waves approach shallower water, they lose up to 60 percent of their energy through friction with the seafloor.

Wave power is intrinsically more difficult to harness than the energy of a flowing river. Waves do not always move in a uniform direction, and any device designed to extract energy under "normal" sea conditions must be able to withstand periodic battering by the most violent storms. These difficulties have not stopped engineers from devising many ingenious devices to put wave energy to work. These devices fall into two broad classes – floating and fixed.

Floating devices are attractive because they can be deployed well offshore where waves are still full of energy, but this also makes them difficult to reach for routine maintenance, and power transmission cables to the shore are long and prone to damage. One promising floating device, nicknamed "the clam," is a doughnut-shaped ring of 12 interconnected chambers, with an overall diameter of about 230ft (70m). Each sealed, air-filled chamber has a flexible rubber wall

facing the open sea. As waves pound against the wall, air within the chamber is forced – via a turbine – into the adjacent chamber, and small generators linked to the turbines generate the power.

Fixed devices are firmly anchored to the seabed in shallow water, or built into the shore, and thus provide better access and reliability. In many designs, wave power is used to compress air, which in turn drives a turbine.

To date, floating devices have been tested only at prototype stage. Fixed devices, however, are already providing power in countries such as Britain, Portugal, and Norway, which are bordered by rough seas and where the high costs of development and installation may have a long-term payback. The World Energy Council estimates that 2TW of power – one-sixth of world demand – could theoretically be harvested from the world's oceans, but the reality is that wave energy conversion is still in its infancy. It is likely that the technology will only gain a foothold over the next few decades in remote and stormy places where conventional energy is expensive.

a turning tide

The daily rise and fall of tides offers an immense, and as yet largely untapped, supply of clean energy – up to 1,000GW of capacity according to some estimates. Tidal mills were used in the Middle Ages throughout northern

wave turbines
Underwater turbines, each about 70ft (20m) in diameter, can produce about 0.6 MW of power from tidal flows. Arrays like this could become a feature of our coasts within a decade.

moon and tides
The rise and fall of the tide is a result of the Moon's gravitational pull on the water in the world's oceans. This force pulls up at the water so that it forms a bulge on the side nearest the Moon. As the Earth rotates daily on its axis, this bulge "passes over" the world's seas around the globe.

Europe to grind corn, but today's plans to exploit the flow of tidal water are on far larger scales. There are basically two engineering approaches to using tides as a renewable energy source. The first employs arrays of large turbines, like underwater windmills, sited in waters where tidal currents are strong. These locations are relatively rare – channels between islands, or around headlands – but there are enough of them to make a significant contribution to future energy needs, and the technology is receiving enthusiastic support from governments and industry.

the severn project

The west coast of England is an ideal location for tidal power stations because of the great daily variation between low and high tide – about 20ft (6m). The Severn Estuary is a particularly good site because the daily ebb and flow of the tides is amplified by the shape of the channel – rather like the sloshing effect when you move back and forward in the bath at just the right frequency. A project proposed in the 1980s envisaged a 10 mile (16km) barrage across the Severn capable of generating about 7 percent of the UK's power demand from 216 turbines along its length. Even back in 1988, the project was costed out at over £8 billion. Construction of the vast barrage was never approved.

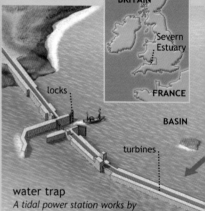

water trap
A tidal power station works by allowing in the advancing tide through sluices, then trapping the tidal water behind the barrage. At low tide, the trapped water is released via sets of turbines linked to electrical generators. The basin of the Severn Estuary barrage would have covered an area of more than 180 square miles (480km²).

The second approach has much in common with existing hydroelectric power stations because it uses vast dam-like barrages to trap incoming tides and then release the trapped water through sets of turbines. Successful pilot schemes have been demonstrated in Russia and France, where the La Rance plant in Brittany has been putting out 240MW of power for the last 30 years. However, the largest and most ambitious projects, such as the 9GW barrage across the Severn Estuary, England, remain on the drawing board, troubled by uncertainties over costs.

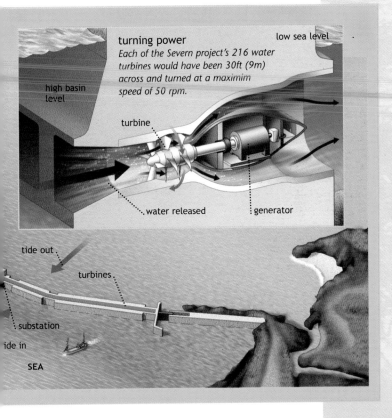

turning power
Each of the Severn project's 216 water turbines would have been 30ft (9m) across and turned at a maximim speed of 50 rpm.

low sea level

high basin level

turbine

water released

generator

tide out

turbines

substation

tide in

SEA

ocean thermal energy

The sea is a giant collector of solar energy. In tropical regions, sunlight warms the surface waters to more than 68°F (25°C), while the waters in the depths remain at an icy 41°F (5°C). A device called an Ocean Thermal Energy Converter (OTEC) can exploit the temperature difference, using it to generate electricity, like a refrigerator in reverse. Prototype OTEC devices consist of a floating unit containing the "fridge" mechanism coupled to a generator. Warm water is drawn from the surface, and cold water is piped up from depths of 0.62 miles (1km). While OTEC is viable in theory, it presents many engineering challenges.

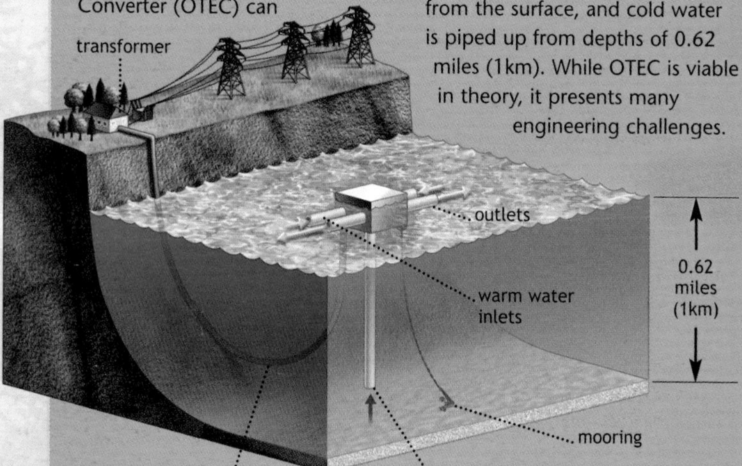

transformer

outlets

warm water inlets

0.62 miles (1km)

mooring

transmission cable

cold water inlet

The prospect of building barrages across the world's largest estuaries has also alarmed environmentalists. The structures would prevent the migration of fish and could affect species of birds that feed on tidal mudflats. On the other hand, barrages would deliver many benefits. Aside from generating plentiful, clean electricity, they would provide protection against the floods that are becoming increasingly common as a consequence of global warming.

farming the wind

Wind power has come of age. It is now actually cheaper to generate electricity from the wind than it is from coal- or nuclear-fueled power stations, and advances in technology are pulling the expanding wind industry into direct cost-competition with gas. Wind energy already supplies 17,000MW worldwide – enough electricity for more than 10 million households – and, according to a leading German investment bank, capacity is likely to grow by 25 percent per annum over the next few years.

Wind is one of the most abundant resources on our planet. It arises when one part of the Earth's atmosphere is heated more by the Sun than an adjacent area is. Differences in air temperature cause differences in air pressure, and a wind is simply the flow of air from an area of high to an area of low pressure. Some winds result from the large-scale circulation of air in the Earth's atmosphere: these prevailing winds blow from the west in temperate

weather front
Hurricane Andrew passes over the Gulf of Mexico in 1992, blasting the area with 155mph (250km/h) winds. Extreme storms like this are rare and capable of causing immense damage, but weaker, more reliable winds can be put to work to produce energy.

climates, and from the east in the tropics, where they are known as trade winds (from an archaic use of the word trade, meaning a course or track). In addition to these main global wind systems, there are predictable local winds – sea breezes and mountain-valley winds – that can be harnessed with the right technology.

People have been using the motive power of the wind for millennia. The first windmills borrowed the design of sailboats, employing giant canvas sails attached to radial arms to turn a central axis. Such devices were used in

Babylon and China from about 2000 BC to pump water and grind grain, and by the 12th century, the technology had spread throughout Europe. In the 1700s, the landscape of pre-industrial England was peppered with more than 10,000 wind machines, but wind power fell out of favor in the ages of steam and coal. It remained in the doldrums until the energy crisis of 1973 prompted new research into commercial-scale wind energy.

turbines
Wind turbines with just two or three blades are far more efficient at high wind speeds than more traditional multivane designs. The winglike airfoil shape of the blades is apparent in this photograph, which shows the installation of a megawatt turbine.

on the wing

Modern wind turbines are inspired more by aerospace technology than by sailboat design – indeed some of the early development work was carried out by NASA for the US government. The first mass-produced, high-efficiency, modern wind turbines appeared in Denmark in the early 1980s. Engineers have achieved substantial improvements in power output and reliability, but the basic design of most of the 20,000 or so wind turbines used worldwide to generate power on a large scale is similar to that of the early Danish models. Two, or sometimes three, winglike blades are attached to a horizontal shaft, which turns an electrical generator via sets of gears. The blades themselves are airfoils that transduce wind energy into motion in the same way that an aircraft's wings generate lift. The entire assembly of blades, gears, and generator can rotate to face squarely into the wind for maximum power conversion.

The power that can be generated by a turbine depends on the diameter of its blades and on windspeed. For this reason, turbines are mounted on towers up to 166ft (50m) tall to maximize exposure to the wind, and are sited on exposed peaks and coastlines. The largest units with huge blades are rated in excess of 1MW, but many smaller units (each producing about 200kW) are often grouped together in their tens or hundreds on "wind farms," which are fast becoming a feature of landscapes across the world. Europe leads the way in this growing industry – 13 percent of Denmark's electricity comes from wind sources – but the US has the most ambitious projects. Plans for a wind farm of 300 turbines at the Nevada Test Site near the California border will produce over 250MW of power by 2005.

farm noise
Critics of wind energy complain of the visual intrusion and noise pollution of wind farms. Siting groups of turbines in remote areas and even offshore has helped to allay these concerns.

making light work

The Sun is a giant fusion reactor. Every second it converts about 11 billion lb (5 billion kg) of matter into energy in a reaction that raises its core temperature to 78 million °F (40 million °C), and the temperature of its surface to 11,000°F (6,000°C). Of course, only a tiny fraction of the Sun's energy output reaches our planet. Most is radiated out into space, or is absorbed or reflected back by the Earth's outer atmosphere. But even with these losses, the power input over just one square yard of the Earth's surface on a sunny summer's day is approximately 1kW. The amount of solar energy falling on the United States over one year is more than 2,000 times the energy produced by all of the nation's coal-fired power stations.

brighter futures

Many scientists believe that solar power will become the single most important renewable technology over the long term. Part of the reason is its ubiquity. Unlike wind, wave, or tidal energy, which can only be usefully exploited in favorable locations, sunlight is everywhere. It can even be put to work in temperate regions persistently covered with cloud. And sunlight can be tapped with small, domestic-scale devices, thus freeing end users from a dependence on centralized power generation.

There are many different means of transducing solar power into more useful forms of heat or electrical energy. At the most basic level are passive systems that use clever design to maximize the amount of "raw" solar energy collected and retained by homes, offices, and factories (see p.31). Active systems are different because they not only collect, but also concentrate and process sunlight to convert it into a higher grade of energy.

One of the first mentions of an active solar system – albeit one with an unusual application – is recorded in Greek legend from the second century BC. The Greek mathematician Archimedes had lent his ingenuity to the Greek struggle against the Romans, who were besieging his home city of Syracuse. Archimedes instructed the Greek soldiers to use their polished bronze shields as mirrors to direct the fierce Mediterranean Sun onto the Romans' boats. In this way, the Greeks were able to set the Roman sails alight before their boats reached shore. The story of Archimedes may well be apocryphal, but it is based on

> **"The use of solar energy has not been opened up because the oil industry does not own the Sun."**
>
> Ralph Nader, politician

solar farm
This solar power station at Barstow, California, consists of a circular pattern of 1,800 mirrors, each measuring 23x23ft (7x7m). The mirrors focus sunlight on to a central collector, heating it to more than 960°F (510°C).

sound scientific principles. The same principles have been used by 20th century engineers to construct giant furnaces and solar power stations in countries such as Spain and Italy, and in California, where sunshine is reliable. These power stations use hundreds of flat mirrors, or heliostats, to direct sunlight onto a central receiver – usually mounted on a tower – and raise its temperature to more than 1,100°F (600°C). Synthetic oils are used to cool the receiver, and carry away the concentrated solar energy, which is then used to generate steam. This in turn drives conventional turbines linked to a generator.

The largest of these solar power stations can produce in excess of 10MW and make a substantial contribution to national electricity needs, but similar technology on a

far smaller scale also has promising applications. In the developing world, for example, as much as 75 percent of the energy used by a household goes to cook food (compared to about one percent in the

US). Gathering firewood is a labor-intensive job, and growing demand for fuel contributes to environmental problems, such as deforestation and desertification. In such countries, the solar stove may provide an answer. The most efficient of these solar stoves are based on a parabolic mirror about 3ft (1m) in diameter. The dish-shaped mirror focuses the sunlight gathered into a small point, where the cooking pot is suspended. Properly used, these solar furnaces can boil a quart of water in less than three minutes. What's more, they are cheap to manufacture and require no fuel, other than sunlight.

in hot water

At present, solar energy is most commonly used in the developed world to provide hot water for homes and swimming pools. In the United States alone, more than two million homes are equipped with "flat plate collectors" that convert sunlight into useful hot water with an efficiency of about 50 percent. The collectors consist of thin metal plates, painted black to maximize

flat plate collectors
Solar water heating units are increasingly installed as standard in the rooftops of new homes. The whole unit is sealed in an insulated box, topped with a double layer of glass that acts like a greenhouse to trap the Sun's heat energy.

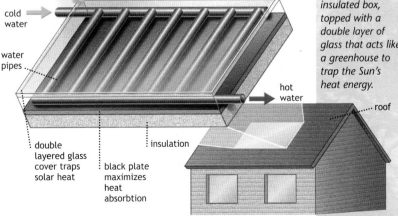

sun

cold water

water pipes

hot water

roof

double layered glass cover traps solar heat

black plate maximizes heat absorbtion

insulation

photovoltaic cells

Photovoltaic (PV) cells are a compact, durable, but expensive source of power. Most cells are made from crystals of the element silicon that has been "doped" with impurites to change the way it conducts electricity. Two distinct types of doped silicon – called n (negative) and p (positive) – are pressed together in a thin wafer and sandwiched beween metal contacts to make a PV cell. When light hits a silicon atom in the p section, it knocks an electron out of the crystal lattice. Impurities in the p silicon prevent the electron from "falling back" into place. Instead, it follows the path of least resistance, migrating over into the n section. The positively charged "hole" left behind by the departed electron in turn migrates into the p section. As more light hits the cell, many electrons build up in the n section, and many holes in the p section. Linking the two together causes an electric current to flow.

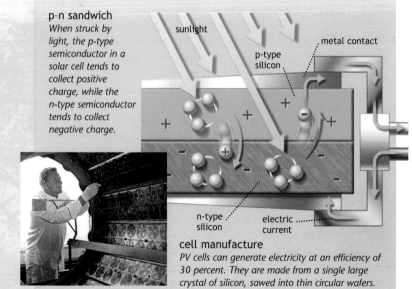

p-n sandwich
When struck by light, the p-type semiconductor in a solar cell tends to collect positive charge, while the n-type semiconductor tends to collect negative charge.

sunlight

metal contact

p-type silicon

n-type silicon

electric current

cell manufacture
PV cells can generate electricity at an efficiency of 30 percent. They are made from a single large crystal of silicon, sawed into thin circular wafers.

absorption of solar radiation. The heat absorbed is transferred to a network of water-filled tubes pressed against, or into, the surface of each plate. The rate of water flow through the tubes is regulated by a thermostatically controlled pump, which ensures that the water is consistently heated to 180°F (82°C), regardless of how strongly the Sun shines. The coolant water then transfers the heat to the home's domestic hot water tank.

A typical home in a sunny state like Florida needs about 40 square feet (4 square meters) of solar collectors to meet its hot water needs, but the beauty of solar heating is that it can be combined with a conventional domestic heating system to cut fuel requirements, even in cooler climates.

built-in power
Photovoltaic panels incorporated into the exterior of this office building in Zurich, Switzerland, generate electricity to supplement the household current supply.

high-tech solar

From the second half of the 20th century, the rising star of renewable energy has undoubtedly been photovoltaic power. Its origins are bathed in the glamour of the space race, when engineers sought new ways of providing reliable power for satellites, but today photovoltaic cells are commonplace, driving everything from pocket calculators to multi-megawatt power stations. The potential of the technology was most spectacularly demonstrated in 1990, when a solar-powered aircraft, the Sun Seeker, flew 2,500 miles (4,060km) across the US, setting a record for fuel-less flight.

Photovoltaic (PV) power is fundamentally different from the other technologies described in this chapter because it depends on a physical principle called the photoelectric effect, discovered by the German physicist Heinrich Hertz in the late 19th century. Hertz noticed

sun

solar panels

microwaves

earth

that certain metals would emit electrons when struck by sunlight, and, with the correct wiring, an electrical current could be made to flow. Photovoltaic cells exploit this effect, generating electricity directly from sunlight, without moving parts, and with no need to create, store, or convert heat energy. What's more, PV cells are made from silicon, which, next to oxygen, is the most abundant element in the Earth's crust. PV technology seems too good to be true, but there is one simple reason that it hasn't supplanted all other forms of energy: cost.

The price of solar cells has fallen by a factor of more than 1,000 since the 1950s, and new breakthroughs in manufacturing continue to reduce the cost. But in cloudy countries, electricity produced by large-scale PV plants is still about eight times more expensive than that from gas-fired power stations. A house in a temperate area needs about 110ft^2 (10m^2) of PV cells to be self-sufficient for electricity, and the cost of between \$10–20,000 currently puts the technology beyond the reach of the majority of domestic users.

However, there are many applications where PV power is ideal. Electricity from solar cells powers radio repeater stations in remote areas, and in the developing world, where two billion people have no access to an electricity grid, photovoltaic cells are a real lifeline. They drive water pumps, lighting systems, and refrigeration units in hospitals, as well as making electronic communication a possibility in isolated villages.

sun satellite

One of the most ambitious plans to harness solar energy involves the launch of a huge PV array into geostationary orbit. The array would beam energy back to Earth in the form of microwaves. Its advantage over Earth-based PV power stations is that it could generate energy 24 hours per day, interrupted only by the occasional solar eclipse.

green generators

Think of a device that collects and stores solar energy, produces no pollution, costs nothing to build, and renews itself throughout its working life. Think of a green plant.

Plants take the raw materials of water from the soil and carbon dioxide from the atmosphere and turn them into oxygen and sugar using energy from sunlight to power the process. Their leaves, stems, and roots are effectively stores of chemical energy, and this can be released when the plant is burned, dies and decomposes, or is eaten by an animal.

Of course, people have been using wood and other biologically derived material (or biomass) as fuel for millennia, and it still accounts for over 85 percent of the energy used in countries such as Kenya and Nepal. Today, the attention of the developed world has also turned to biomass because it provides a real alternative to our dependence on fossil fuels. Biomass, or biofuels, currently provide about 3.6 percent of the energy used in the US, and new European targets for renewable energy will see biofuels contributing 8.3 percent of energy consumption by 2010. Biofuels take many, diverse forms. Some are crops, such as willow coppice and *Miscanthus* (a bamboolike grass), which are grown specially for their energy content. These plants are used because they grow

grown to burn
Willow coppice, grown specially as an energy crop, is harvested and burned in a furnace just four years after planting.

where there's muck

Every year, Americans throw out 84 million tons of paper and 44 million tons of food waste. In New York State alone, enough garbage is produced annually to cover a football field to a height of 3 miles (5km). Some of this waste is incinerated, and the heat produced used to generate electricity; some is recycled; but a large portion is simply buried in giant, plastic-lined holes in the ground – landfill sites. Over the years, this dumped waste rots down, releasing a mixture of gases – mainly methane and carbon dioxide. The gas can be a hazard if new buildings are erected near or on an old landfill site, but it can also be put to good use. Many modern landfill sites are peppered with a network of perforated pipes buried at depths of up to 70ft (20m) in the refuse. These pipes collect the combustible gas and carry it to storage areas. From there, it is piped off to be burned in a power station to produce electricity for the national grid.

waste mountain
Garbage is dumped into the Fresh Kills landfill site near New York City. This is the largest landfill in the world, and is officially the tallest "mountain" on the East Coast of the US.

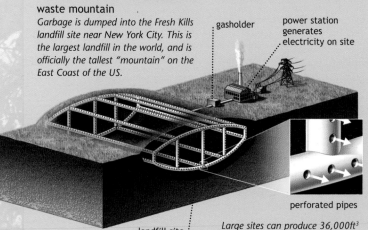

gasholder

power station generates electricity on site

perforated pipes

landfill site

Large sites can produce 36,000ft³ (1,000m³) of gas per hour.

rapidly and can be harvested using modified farm equipment. They have the added benefit that they can be grown on surplus agricultural land – and they also provide new habitats for wildlife in the bargain.

Other biofuels are the byproducts of farming and forestry: with the correct processing, straw, wood chippings, rice husks, coconut fiber, domestic garbage, and animal wastes, such as chicken manure, can all be used as fuels.

golden litter
Poultry litter is a form of dry waste that makes an excellent fuel. One power station in the UK produces 12.5MW of power from nothing else.

gas chips
Wood chips are loaded into an experimental reactor that converts biomass into methane-rich gas, which is a good substitute for natural gas.

processing plants

There are three main ways in which biofuels are used. First, solid fuels can be burned in the home to provide heat, or in power stations to generate both heat and electricity. Crucially, when burned, they produce no more carbon dioxide than if they had simply decayed naturally, so their use makes no net contribution to global warming. Second, they can be processed into liquid fuels: bio-ethanol (alcohol) can be made from forestry residues, straw, sugar cane, and maize, by fermentation and distillation. And vegetable oils, such as palm and rapeseed oil, can be processed into substitutes for diesel oil. These concentrated liquid biofuels are easy to transport and can be used to power cars and trucks; indeed, a bio-ethanol mixture was sold at gas stations during the US fuel crises of the 1970s to protect dwindling gas reserves.

biogas plant
This huge installation in northern Italy produces combustible gas from the remains of sorghum crops.

one for the road
More than 40 percent of the fuel used by cars in Brazil is alcohol (ethanol), made by distilling sugarcane.

The economics of ethanol production are still marginal, but ongoing research is sure to unlock the potential of this clean-burning fuel.

Third, biomass can be treated to convert it into a combustible gas. In a technique called gasification, wood is heated under great pressure with a mixture of steam and oxygen. The resulting gas mixture, which has about one-tenth of the energy value of pure methane, can be "scrubbed" to remove pollutants, and then burned in conventional high-efficiency gas turbines (see p.10) to generate electricity. Another way of making an energy-rich gas from biomass is to put bacteria to work. In warm, wet, airless conditions, bacteria will digest any organic material, producing combustible methane gas as a by-product. Commercial "biogas" production is already big business. The raw materials – usually animal waste or wet plant matter – are fed into a large metal reactor with a capacity of up to 72,000ft³ (2,000m³) and maintained at a temperature of 95°F (35°C) for 10–25 days. The gas produced – about 14,000ft³ (400m³) for every dry ton of raw material – is tapped off and burned in a power station to generate electricity, while the remaining sludge is used as an agricultural fertilizer.

Domestic-scale biogas is already an important source of power in developing countries, with two-thirds of China's rural households relying on biogas as their primary fuel. The technology is fast catching on in the West, where biogas units that use household waste are providing renewable power at the local level.

energy from the earth

We are all standing on the surface of a giant boiler. Thousands of miles beneath our feet, energy released by the decay of naturally occurring radioactive elements keeps the interior of our planet at temperatures of up to 12,600°F (7,000°C). This vast reservoir of heat is spectacularly evident when molten rock (magma) erupts volcanically through a flaw in the Earth's solid crust, and wherever hot water and steam emerge to the surface as hot springs and geysers.

People have been exploiting free energy from the Earth (geothermal energy) for centuries. The Romans piped steam from underground to heat their homes, and geothermal energy is used to heat houses, offices, and factories in many cities, such as Reykjavik, Iceland.

eruption
Energy deep below the Earth's crust is released in the eruption of Kilauea, Hawaii, one of the world's most active volcanoes. The temperature of the newly erupted lava can exceed 2,200°F (1,200°C).

crust
outer mantle
inner mantle
liquid outer core
solid inner core

radius 3,900 miles (6,370km)

The real potential of geothermal energy was unlocked when underground steam was used to generate electricity. The first geothermal power station was commissioned in 1913 at Larderello, northern Italy, and today geothermal energy is among the most promising of all renewable resources. In the US, it accounts for 2,850MW of power generation – almost four times as much as wind and solar energy combined – and the world's largest geothermal power station at The Geysers, California, puts out enough power to supply the cities of San Francisco and Oakland.

Heat from within the Earth rises up across the whole surface of the planet, but there are relatively few places where it is sufficiently concentrated to be harvested on an

molten planet
The Earth is a layered structure, and its temperature increases with depth. Within the solid crust, temperature rises by about 5°F (2°C) every 300ft (90m). At the top of the mantle, the semi-molten rock is at about 1,830°F (1,000°C). The temperature then increases slowly toward the core. Only in some places does molten magma push up through cracks in the crust, nearing the surface to create "hot spots."

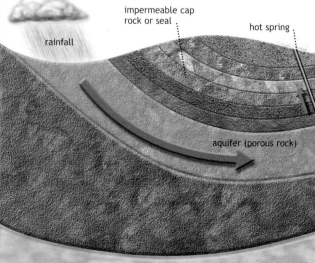

impermeable cap rock or seal
hot spring
rainfall
aquifer (porous rock)

economic scale. These places tend to be wherever the semi-liquid rock of the mantle rises up through the Earth's crust to within one mile (1.6km) of the surface to create a "hot spot."

It is possible to harvest heat by drilling through the crust into the magma, and pumping water through the hot, molten rock to extract its energy, but this head-on approach is rather dangerous because lava can erupt through the boreholes. In practice, most commercial geothermal energy is extracted from groundwater that has been heated to between 300°F (150°C) and 480°F (250°C) and vaporized into steam by underlying magma. This steam is used to drive turbines and generators as in a conventional power station (see p.10). If a convenient

the ring of fire
Hot molten rock nears the Earth's surface wherever the plates of the Earth's crust meet (red). These areas – high in potential for geothermal energy – are also major sites of volcanoes and earthquakes.

earth generator
Geothermal energy can be tapped wherever molten rock nears the Earth's surface and heats water in an aquifer. The trapped steam is extracted through boreholes and used to drive turbines. Steam and hot water also escape naturally from aquifers via hot springs, geysers, and fumaroles.

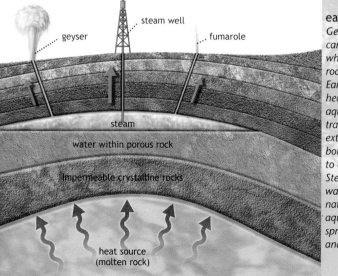

geyser
steam well
fumarole
steam
water within porous rock
impermeable crystalline rocks
heat source (molten rock)

geothermal power station
The warm waters of the Blue Lagoon, next to the Svartsengi power station in Iceland, are constantly replenished by hot wastewater from the power plant.

aquifer is not present above a hot spot, an artificial underground reservoir can be created by pumping water into fissures in the rock; piping steam back up to the surface extracts heat energy from the depths.

prospecting the depths

What makes geothermal energy such an attractive source of renewable power is its concentration. Unlike wind, wave, and solar energy, which are diffuse, geothermal energy can be harvested economically from a point source. Globally, the geothermal generation industry is growing at about eight percent a year and best estimates suggest that it could one day supply 115,000MW of power in the US alone. Yet there are voices of dissent: environmentalists point out that subterranean steam carries with it to the surface pollutants such as hydrogen sulfide and toxic minerals, which later reach streams, rivers, and lakes, where they can affect aquatic life. Critics also point out that geothermal plants also emit the greenhouse gas carbon dioxide. However, the emissions are at very low levels – about one-thousandth the amount of a fossil fuel power station of equivalent capacity.

alternative realities

Smart investors know that green energy is set to be one of the key growth industries in the 21st century. Globally, wind power alone is already worth more than $2.3 billion, and the leading players in the energy market are hurriedly diversifying into renewables. Shell estimates that 50 percent of the world's energy needs could be met by renewables by 2050. In the past, the main argument against renewables has been one of cost: why research and develop new sources of energy when fossil fuels are cheap and abundant? But this argument begins to appear hollow when the true costs of fossil fuel use are revealed. The International Center for Technology Assessment, a leading US think tank, calculates that $1-worth of gas bought at the filling station actually costs between $5 and $15 when we take into account hidden costs, including the costs of air, land, and sea pollution, and the generous tax breaks

new and old
In the immediate future, our energy needs will be met by a mixture of conventional fossil fuels and renewables. This gas station in west London uses photovoltaic panels to convert sunlight into electricity, which then powers the gas pumps and lighting.

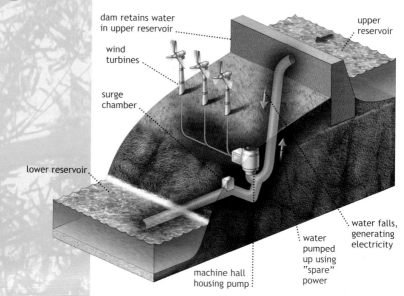

dam retains water in upper reservoir

upper reservoir

wind turbines

surge chamber

lower reservoir

water falls, generating electricity

water pumped up using "spare" power

machine hall housing pump

pumped storage

The energy output of a wind farm can be used to power a pump that moves water from one reservoir to another, higher reservoir. The stored energy can be released on demand simply by reversing the flow and using the pump as a generator that feeds into the electricity grid.

bestowed on the oil companies. Many governments are seeking to level the playing field by introducing fairer tax structures that reward cleaner technologies, by directly funding renewable technologies, and by setting green energy targets. In the UK, for example, power companies will be legally obliged to produce 10 percent of their electricity output from renewable sources by 2010. Consumers are also driving the move to renewables. In the late 1990s, deregulation in the energy business meant that people could choose which company supplied the electricity to their home. Many opted for companies with the best environmental credentials, and today there are numerous energy schemes that supply "green" electricity from renewable sources.

Over the next few years, renewables will increasingly supplement fossil fuels and nuclear energy; in the longer term they could – theoretically – supplant conventional sources of power. If this is to happen, one problem remains to be overcome – predictability of supply. Energy is only useful if it is in the right place, in the right form, at the right time. For example, consumption of electricity from the UK's national grid is lowest at 2:00am, but rises by 70 percent at 11:00am; and far more is used in winter than in the summer months. Gas-fired power stations can be switched on and off to match this demand, but this is not an option with renewables – we cannot control the winds, waves, and tides.

The answer is to store up excess energy from renewable sources, then release it to smooth out fluctuations in demand. This can be done easily by pumping water upward when energy is plentiful, and releasing it in times of demand; but more advanced technology provides a better solution. The excess renewable energy, in the form of electricity or heat, can be used to split water into its constituent elements – hydrogen and oxygen gases. These gases can be liquefied for ease of storage and transport, and then recombined in a device called a fuel cell to produce electricity. Fuel cells work at efficiencies of more than 70 percent, and produce only water as a waste product. They are already being used to drive buses and cars and are hailed by many as *the* power source of the future.

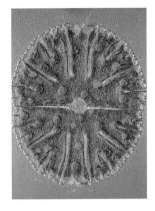

green hydrogen
Under the right conditions, some algae split water to produce hydrogen gas. Scientists at the University of California are assessing the potential of algal hydrogen as a fuel for the future.

clean car
This car has been adapted to run on hydrogen gas in a fuel cell. It can run for 190 miles (300km) on a single tank of hydrogen, and produces no atmospheric pollution.

house of the future

In the coming decades, new energy technologies are likely to be adopted by the world's largest power companies. The electricity supplied to our homes will come increasingly from renewable sources, and we may well be filling our cars with ethanol or hydrogen pumped from the filling station. But the beauty of renewable energy is that it can also be applied on a smaller, domestic scale. While we cannot operate our own personal oil refineries and power stations, we can plug into electricity generated on our rooftops or use the Sun to heat our homes. Countless technological innovations are being built into new homes to maximize their energy efficiency, often supported by government subsidies and tax breaks. As the costs of these technologies fall, and the prices of fossil fuels rise, energy-efficient, or even self-sufficient, homes will become the rule rather than the exception. The diagram here brings together much of what is possible with today's technologies in a single home.

.... wind turbine supplying local area

trees screen the house from prevailing winds, thus cutting heat loss .

eco-house
The Integer house, developed in the UK, showcases many energy-saving features.

hi-tech control
Information technology plays a big part in the home of the future. Computers control power output and internal temperature; and microchips allow everyday devices to communicate to optimize power use.

water in tank is heated by the rooftop solar collectors. A conventional electric element gives extra heating in the coldest conditions

flat plate collectors produce hot water, even in cloudy conditions

photovoltaic panels produce electricity, for storage in battery chamber

deep overhang shades windows from the intense summer sun

thermostatically controlled pump

small windows on north-facing walls

intelligent appliances

thick, insulated walls

battery array plus inverter and transformer provide household current-voltage electricity

large south-facing window fitted with computer-controlled blinds

glossary

acid rain
Rain with a high concentration of sulfuric or nitric acid.

active solar heating
Heating a building using a solar collector connected to pumps or fans that transfer the heat within.

airfoil
A body shaped in such a way that it produces lift in a direction that is at right angles to its direction of motion, or to the direction of the fluid moving past it.

anticline
A large upfold of rock strata, usually 60–180 miles (100–300km) in diameter.

aquifer
A permeable rock formation that stores and transmits groundwater in sufficient quantities to supply wells.

armature
Coiled conducting wire that forms the moving part of an electrical **generator**.

barrage
An artificial obstruction placed in a watercourse to block its flow and increase its depth.

barrel
A unit of capacity often used in the oil industry. One barrel is equivalent to 42 gallons (160 liters).

biogas
A mixture, principally, of **methane** and **carbon dioxide** produced by the fermentation of organic matter.

biomass
Any form of organic material that contains energy stored in chemical form, usually in compounds of the element carbon. Biomass includes animal manure, crop residue, human refuse, and wood.

British thermal unit
A unit of energy. One unit (Btu) represents the amount of energy needed to raise the temperature of 1lb of water by 1°F.

Btu
see **British thermal unit**

carbon dioxide
A colorless gas formed when carbon-based fuels are burned. Naturally present in the Earth's atmosphere, the gas contributes to the greenhouse effect.

chemical energy
The energy stored in the chemical bonds of a molecule.

electrical energy
The energy contained by a stream of electrons flowing in a circuit, or in electromagnetic waves, such as light waves.

electron
A fundamental particle that carries a negative charge. Electrons are a basic constituent of atoms.

ethane
A colorless, odorless, flammable gas that is a common constituent of **natural gas**.

ethanol
A colorless organic liquid that can be formed by the fermentation of sugars. Also known as ethyl alcohol.

first law of thermodynamics
Physical law that sums up the conservation of energy in a closed system. The heat added to a system plus the work done equals the change in the total energy of the system.

fission
The spontaneous or induced splitting of a heavy atomic nucleus into two or more lighter fragments.

fossil fuel
A fuel – coal, oil, or natural gas – formed from the fossilized remains of plants and animals.

fuel cell
A device that produces an electric current from a chemical reaction between hydrogen and oxygen.

fusion
The process of bringing together two atomic nuclei to form one larger nucleus. Energy is released through loss of mass in the product.

gas turbine
A device that produces electricity by harnessing the force of hot, expanding gases – usually produced by burning a fuel – to make a turbine revolve.

generator
A device that converts mechanical energy (movement) into electrical energy.

geothermal power
Power generated by tapping into the heat energy stored naturally in the rocks of the Earth's crust.

giga (G)
Prefix used to denote a multiplier of 1,000,000,000 (10^9). One gigawatt (GW) is 10^9 **watts**.

gravitational force
A fundamental force of nature; the force of attraction between two masses.

green electricity
Colloquial term used to describe electricity that has been generated from **renewable resources**.

greenhouse effect
Warming of the Earth's atmosphere due to the entrapment of heat by gases in the air. These gases are primarily **carbon dioxide**, **methane**, and ozone.

grid
A national or regional network of electrical power generation, transmission, and distribution.

half life
Time taken for half of a sample of a radioisotope to decay.

head
In **hydroelectric** generation, the height of water trapped behind a dam, as measured from the level of the turbines.

heat exchanger
Any device used to transfer heat between two gases or liquids.

hydrocarbon
Any chemical compound that contains only hydrogen and carbon atoms.

hydroelectric
An electric generator for which the mechanical **power** is provided by moving water.

hydrothermal
A type of **geothermal** power in which water heated below the ground in aquifers is used to drive **turbines** on the surface.

insolation
The amount of sunlight energy falling on a given surface area of land.

isotopes
Atoms with the same numbers of **protons** and **electrons** but with different numbers of **neutrons**.

joule
The standard international unit of energy and **work**. One joule (J) is the work done when a force of one **newton** (N) moves its point of application through one meter.

kilo (k)
Prefix used to denote a multiplier of 1,000. One kilowatt (kW) is 1,000 **watts**.

kinetic energy
The energy possessed by moving objects.

light
A form of electromagnetic radiation with a wavelength of between about 400 and 800nm that can be detected by the human eye.

magma
Molten rock under the Earth's crust and mantle. Magma that reaches the surface is called lava.

mega (M)
Prefix used to denote a multiplier of 1,000,000 (10^6). One megawatt (MW) is 10^6 **watts**.

methane
A colorless, odorless gas that is the main constituent of **natural gas**.

natural gas
Any gas found in the Earth's crust, but most often applied to **hydrocarbon** gases associated with petroleum extraction. In this context, natural gas is principally **methane**, **ethane**, and **carbon dioxide**.

neutron

A basic constituent of the atom, located in the **nucleus** and carrying no electric charge.

newton

The standard international unit of force. One newton (N) is the force required to impart an acceleration of $1m/s^2$ to a mass of 1kg.

nucleus

The dense, positively charged central part of an atom, made up of **protons** and **neutrons**.

ore

A naturally occurring rock that holds valuable minerals.

passive solar heating

Utilizing a building itself to collect and store incoming solar radiation. Some buildings are designed specially to maximize their heat-trapping properties.

penstock

In hydroelectric power stations, a high-capacity pipe that carries water to the **turbine**.

petrochemical

Chemical derived from crude oil or **natural gas**.

pH

The measure of the acidity or alkalinity of a chemical. Pure water has a pH of 7; acids have lower figures, alkalis higher.

photovoltaic effect

The emission of an **electron** from the surface of a metal when it is struck by light.

plasma

An electrically conductive gas composed of equal numbers of positive **nuclei** and **electrons**.

potential energy

The energy stored by virtue of an object's position or shape.

proton

A basic constituent of the atom, located in the **nucleus** and carrying a positive charge.

pumped storage

A means of storing excess **electrical energy**. The energy is used to pump water up a gradient; the water is released and used to generate power at times of high demand.

radioactivity

The spontaneous disintegration of certain heavy elements accompanied by the emission of X-rays, fast-moving **electrons**, or positively charged helium **nuclei**.

renewable resource

An energy source that is naturally replenished.

reserve

The amount of an energy resource that is known to be recoverable using current extraction techniques and technologies.

short ton

Unit of weight often used to measure coal resources. One short ton is equivalent to 2,000lb or 0.907 tonnes.

solar cell

A device with no moving parts that relies on the **photovoltaic effect** to collect solar energy from sunlight and convert it directly into electricity.

sustainable development

Development that meets today's needs while protecting the environment and its resources for future generations.

tera (T)

Prefix used to denote a multiplier of 1,000,000,000,000 (10^{12}). One terawatt (TW) is 10^{12} **watts**.

thermodynamics

The study of energy flow and conversion.

thermostat

A control device that maintains a system at constant temperature.

turbine

A machine composed of a set of blades mounted on a central shaft, which is made to rotate by a moving fluid, such as water, steam, or another gas, usually to turn an electrical **generator**.

watt

The standard international unit of power. One watt (W) is equivalent to a power output or use of one **joule** per second.

work

A measure of the change in energy when a force causes an object to move.

index